Abel Hernández-Muñoz

Cats, enigmatic friends

Abel Hernández-Muñoz

Cats, enigmatic friends

Introductory guide to the breeding of these felines

ScienciaScripts

Imprint

Any brand names and product names mentioned in this book are subject to trademark, brand or patent protection and are trademarks or registered trademarks of their respective holders. The use of brand names, product names, common names, trade names, product descriptions etc. even without a particular marking in this work is in no way to be construed to mean that such names may be regarded as unrestricted in respect of trademark and brand protection legislation and could thus be used by anyone.

Cover image: www.ingimage.com

This book is a translation from the original published under ISBN 978-613-9-43800-6.

Publisher:
Sciencia Scripts
is a trademark of
Dodo Books Indian Ocean Ltd. and OmniScriptum S.R.L publishing group

120 High Road, East Finchley, London, N2 9ED, United Kingdom
Str. Armeneasca 28/1, office 1, Chisinau MD-2012, Republic of Moldova, Europe
Printed at: see last page
ISBN: 978-620-8-15240-6

CATS, ENIGMATIC FRIENDS
Introductory guide to the breeding of these felines

MSc. Abel Hernández Muñoz

Table of Contents

INTRODUCTION

Humans domesticated some animals for food, clothing, work, and as pets or companion animals. How this happened is controversial. Through protection and selective breeding, humans transformed the first domesticated animals into more productive breeds, such as cattle, sheep and poultry. Also contributing to human welfare are dogs, cats, white rats and mice, guinea pigs and monkeys that medical research has used to increase knowledge of human physiology and to develop drugs and procedures to combat human diseases.

However, as our species continues to spread across the Earth, it invades and contaminates the environments of many animals, reducing the remaining habitats to smaller and smaller areas. Unless this trend is reversed, most animal life faces extinction.

The cat became man's companion rather late, about 4,000 years ago in Egypt. It is not clear why the Egyptians tamed the wild cat. They may have been tamed for practical reasons, as the cat killed rodent pests that plagued grain stores, or for religious reasons, as the cat played an important role in Egyptian religion as a representation of the goddess Bastet.

The wild cat, unlike the ancestors of all other major domestic animals, is not sociable. It is not difficult to imagine how human groups could have acted as packs, herds, flocks or flocks for wolf cubs, wild boars, sheep or wild calves, but what is not so easy is to know why a wild cat would agree to have contact with humans, if not to secure food.

The solitary feral cat caused its domestic offspring, whose behavior is reminiscent of their ancestors in many respects, to inherit that tendency to be independent. In fact, except for some unimportant changes in color, build and size, most domestic cats have a physical appearance that resembles their wild ancestors to an unsuspected degree.

Today they live all over the world and delight millions of people who enjoy them as pets. In addition to these companion cats, there are also working cats that eliminate unwanted animals on farms, ranches and in towns. Although bad-tempered, both feral and house cats can reduce colonies of birds and small mammals.

GENERAL INFORMATION ABOUT THE CAT

1 GENERAL

Cat, small, mainly carnivorous animal *(Felis catus),* belonging to the Felidae family. Popular as a domestic animal and prized as a hunter of mice and rats. Like almost all members of the feline family, the domestic cat has retractile nails, good hearing and smell, remarkable night vision and a compact, muscular and very flexible body. The cat has an excellent memory and shows considerable aptitude for learning by observation and experience. The natural life expectancy of the domestic cat is about 15 years.

2 ORIGIN OF SPECIES

Most scientists believe that the short-haired varieties of the domestic cat are derived from the cat *Felis libyca,* a species of African wild cat domesticated by the ancient Egyptians, perhaps as early as 2500 BC, and transported by the knights of the Crusades to Europe, where they interbred with the smaller native wild cats. The long-haired breeds may be descended from the Asian wild cat *(Felis manul).* Over the centuries, the cat has maintained practically the same size, weighing approximately 4 kg at the completion of its development, and has preserved its instinct for solitary hunting.

2.1 Cat physiology

The body of a domestic cat is extremely flexible: its skeleton is made up of more than 230 bones (the human skeleton, although much larger, contains only 206 bones), its pelvis and shoulders are attached to the spine much more loosely than in most quadrupeds. The cat's great jumping ability is due, in part, to its powerful musculature. The tail provides stability when jumping or falling.

The cat's claws are designed to capture and hold prey. The sharp, curved, retractable claws are sheathed in a soft, leathery pad at the end of each toe, and are brought out to fight, hunt or climb. The cat marks its territory by scratching and leaving its scent on trees or other objects; its claws leave visible scratches and the scent glands in the pads leave their scent.

A cat's teeth are for biting, not chewing. Its powerful jaw muscles and sharp teeth allow it to take a deadly bite out of its prey.

2.2 Senses

The cat's eyesight is exceptionally adapted to hunting, especially at night. It has excellent night vision, very wide peripheral vision and binocular vision that allows it to judge distances accurately. Cat day vision is not as good as that of humans; cats see movement much more easily than detail and are thought to see only a limited range of colors.

The cat has extremely sensitive hearing. It can hear a wide range of sounds, including ultrasonic sounds. Its auditory sense is less sensitive to low frequencies, which may explain why some domestic cats are more receptive to female voices than to male voices. The cat turns its ears independently to focus on different sounds.

The cat's sense of smell is highly developed and plays a vital role in foraging and reproduction. Many of the social signals of domestic cats take the form of scent: for example, males can apparently smell a female in heat from hundreds of meters away.

The cat has a specialized sense of taste in a peculiar way: it has little ability to detect sweetness, but is very sensitive to slight variations in the taste of water. The cat's tongue is covered with rough bumps, or *paps,* which it uses to scrape meat from bones. It also uses its tongue to clean itself.

The whiskers or vibrissae are very sensitive to the slightest touch and are used to warn of obstacles and notice changes in the environment; in low light they are used to find the way.

2.3 Reproduction

The domestic cat reaches puberty at about nine or ten months of age. A sexually mature female cat goes into estrus, or *estrus,* several times a year; during estrus she is both receptive and attractive to cats. The gestation period is about 65 days and the usual litter of 4 kittens. The kittens are born deaf, blind and helpless; they open their eyes 8 to 10 days after birth and weaning begins at 6 weeks of age.

2.4 Coat colors

The original coat color of the domestic cat was probably grayish-brown with darker patches, a color that provides excellent camouflage in various environments. All other colors and patterns are the result of genetic mutations; for example, solid-colored coats, such as black or blue, are due to a gene that suppresses stripes; the reddish coat to a gene that transforms black pigment to reddish, and the white coat is the result of a gene that completely suppresses all pigment formation.

Two pigments, black and orange, form the basis of all modern domestic cat colorations. These pigments can be combined with each other or with white (no pigment). A single gene, the O gene (from *orange*), determines whether a cat's coat contains orange or black pigment. We can imagine the O gene as a switch that is either on (orange pigment) or off (black pigment). This gene is located on the X chromosome, so its inheritance is sex-related.

3 CAT BREEDS

There are about 40 varieties or breeds of domestic cat recognized worldwide. Although the various breeds differ radically in tail length and general appearance, they vary less in size than dog breeds. The smallest breeds weigh about 2 to 3 kg when the cat is an adult and the larger breeds weigh 7 to 9 kg. Attempts to develop miniature or giant domestic cats have so far failed.

3.1 Origins of the breeds

Many domestic breeds, including the Maine Coon, Manx, Russian Blue and Siamese originated as a natural variety of the domestic cat specific to a geographic area. Others, such as the Himalayan, are human-created breeds, the result of careful breeding over generations to achieve the desired appearance. Some relatively new breeds, such as the rex (curly-haired), sphynx (hairless), scottish fold (with folded ears) and american curl (with ears covered with curly hair), arose from genetic mutation and were developed as a distinct breed through selective breeding.

3.2 Breed standards

For each breed of domestic cat there is a model of perfection, declared by different associations of cat owners, which describes the ideal cat of a given breed and its distinctive features, defines the ideal and non-ideal characteristics and mentions the defects which, in a cat show, could be grounds for penalties or disqualification. For example, in the Siamese cat standard the eyes are described as almond-shaped and slanting towards the muzzle and squinting is a disqualifying defect.

The models of each breed differ slightly from one association to another and not all associations recognize all breeds. To be accepted into an association, a breed must first be provisionally accepted. In order to participate in championships, the breed must meet a series of requirements that vary according to the association.

Modern cat breeds

BREED	LENGTH OF HAIR	DESCRIPTION
Abyssinian	Short	A graceful and stylized cat, with a ticked somewhat similar to that of the wild rabbit.
American curl	Semi-long	New breed, with gently upward curving ears, semi-long coat, and medium to large hazel eyes.
American curl	Short	Short hair version of the American curl.
e short hair		
American shorthair	Short	Cat native to North America, short-haired and good at hunting rodents. It is a robust animal with an elongated nose, which exists in a wide variety of colors.
American wirehair	Short	Similar to the American shorthair, except for its coat, which is thick, woolly and coarse. Its whiskers are curved and can project at odd angles.
Balinese	Semi-long	Long-haired version of the Siamese. Semi-long, fine, silky and variegated coat.
Bengal	Short	A large cat with a golden-brown spotted coat that resembles a small leopard. The Bengal breed arose from the crossbreeding of the domestic cat with the Asian leopard cat.
Sacred Burma	of Semilargo	Cat shrouded in mystery and legend. Intense blue eyes, coat with even-toned ends and pure white gloves.
Bombay	Short	Medium-sized muscular cat with an attractively rounded head. It is characterized by its shiny 'panther' coat and large, bright, intense copper-colored eyes.
British shorthair	Short	Cat native to Great Britain, very useful for rodent control. Strong, massive and robust, it has a round head and eyes, and a short, dense and frizzy coat.
Burmese	Short	Strong and compact cat, with a round head and large round golden eyes. Its short, satin touch coat comes in continuous jet, champagne, blue and red colors.
California spangled	Short	Recent breed, with spots or rosettes on the back and flanks, and stripes on the neck and shoulders.
Chartreux	Short	Ancient French breed. With a robust and agile body, it has a wide and smiling head. Its short and dense coat can have a bluish hue.
European colorpoint	Short	Any Siamese other than point seal, blue, chocolate or lilac.
Cornish rex	Short	Short, wavy coat, very large ears and long, slender body, with a sharp nose like a greyhound.

Breed	Coat	Description
Cymric	Long	Long-haired version of the manx.
Devon Rex	Short	Elf face and ears, and soft, wavy fur.
Egyptian Mau	Short	The only natural breed of spotted cat. Almond-shaped gooseberry-colored eyes and short, spotted coat.
Exotic shorthair	Short	Shorthair version of the Persian. Short, soft and plush coat.
Havana	Short	Medium sized cat with deep green eyes and short, glossy, deep warm brown coat.
Himalayan	Long	Similar to the Persian, from which it differs by its markings, which are points (like the Siamese).
Japanese Bobtail	Short	Cat native to Japan. Long and elegant body, with a short tail ending in a tassel. It comes in a variety of colors. The colorful 'mi-ke' (black, red and white tortoiseshell) is the most popular because it is said to bring good luck.
Javanese	Semi-long	Any Balinese other than point seal, blue, chocolate or lilac.
Kashmir	Long	Chocolate or lilac Persian.
Korat	Short	Native to Thailand, where it is often considered a good luck cat. Heart-shaped face with large round eyes and solid silver-blue fur.
Maine coon	Long	Long-haired cat native to North America. Animal of good size and better character, woolly coat that does not need special care.
Manx	Short	Tailless cat native to the Isle of Man. Short, firm and rounded body, its hind legs are longer than the front ones.
Nibelungo	Long	Long-haired version of Russian blue.
Norwegian of Forest	Largo	Native Norwegian cat, good rodent hunter. Medium to large size, long water-repellent coat, triangular head with tufted ears, large almond-shaped eyes and long bushy tail.
Ocicat	Short	Large spotted breed, originated from the crossbreeding of Abyssinians, Siamese and American Shorthairs.
Oriental	Semi-long	Unblemished version of the Balinese. Green-eyed and of different colors, except for the point.
Oriental short hair	de Corto	Non-spotted version of the Siamese. Green eyes and different colors, except for the point.
Persian	Long	Short and stocky cat, with massive bones, very long and thick fur, short and wide nose, small ears, and large round eyes.
Ragdoll	Semi-long	Large breed whose name ('ragdoll') is due to its docile and very calm temperament. Ragdolls come in various colors, including pointed patterns (Siamese) with or without white mittens on the paws.
Russian blue	Short	Graceful cat with big ears and bright emerald green eyes. His short, double coat is uniform blue with a silver ruff.

Scottish fold	Short	Stubby cat with small, folded ears and a sympathetic expression of sadness. Its dense white coat has a great diversity of patterns and colors.
Scottish fold Long hair	Long long	Long hair version of the scottish fold.
Selkirk rex	Short	Large, muscular cat with a plush, curly coat.
Selkirk rex of longhair	Semilargo	Semi-long-haired version of the selkirk rex.
Siamese	Short	Slender cat with blue eyes and characteristic coloration: creamy white body with darker colors on legs, ears, mask and tail.
Singapore	Short	Small and stocky cat native to Singapore. Large almond-shaped eyes, short brown ticked fur, similar to that of wild rabbits.
Snowshoe	Short	Elongated and stocky cat, with point colors similar to those of the Siamese, white paws and facial mask. The breed was generated by crossing Siamese with two-colored American Shorthairs.
Somali	Semi-long	Longer hair version than the Abyssinian. Semi-long coat and shaggy tail.
Sphynx	Short	Almost hairless cat, with very large ears and a pixie face. The fur has a rough appearance, and to the touch, it is like a soft and warm suede.
Tiffanie	Long	Long-haired version of the Burmese.
Tonkinese	Short	This breed was originally created by crosses between Siamese and Burmese. Its paws, tail, ear and nose have dark colors, while its body is slightly lighter. Blue-green eyes, and fine and shiny coat in cream, lilac, blue and platinum colors.
Turkish angora	Semi-long	Long, slender cat with wedge-shaped head, oval eyes, elegant, fine, silky coat and bushy tail.
Turkish Van	Semi-long	An ancient breed native to the mountainous regions of Eastern Turkey, near Lake Van. It is a predominantly white cat, with one or two colored spots on the head, tail and body. Turkish Van cats are characterized by their love of water.

4 CAT CARE

Cats, like all pets, depend on humans for their care and feeding, and require considerable attention.

4.1 General care

Although cats have a reputation for being relatively independent, domesticated cats need the love and attention of their owners. A balanced daily diet, such as that provided by high-quality cat food sold in stores, is essential to their health and longevity, as is a regular supply of fresh water. Regular cleaning of the litter tray is necessary to prevent disease; some cats do not use it when it is not very clean. Cats should have their nails trimmed frequently. To prevent damage to furniture, cats should be provided with a scratching post, preferably made of a rough material such as pita rope. Cats use their tongues to clean their fur and usually eat all loose hairs. All cats, including shorthaired cats, should be brushed weekly to remove all loose hair, this helps prevent hairball formation in the stomach. Some long-haired breeds, such as the Persian or Himalayan, require daily brushing to keep their long, soft coats from matting.

4.2 Cat diseases

Domestic cats are susceptible to developing a number of viral and bacterial diseases. Fortunately, many feline diseases can be controlled by a regular course of vaccinations. Cats can also suffer from external parasites, such as fleas, roundworms and intestinal parasites (worms).

Respiratory infections are a common disease and can be fatal, especially in young kittens. Vaccines offer some protection against the following respiratory diseases: *feline viral rhinotracheitis* (FVR), *feline plague and feline pneumonia.*

Feline infectious enteritis is a highly contagious, often fatal disease characterized by a sudden attack and various gastrointestinal symptoms such as vomiting or diarrhea. Vaccination is the only effective way to control the disease.

Feline leukemia is a contagious and fatal disease that is transmitted by direct contact. A cat with feline leukemia exhibits various symptoms, including general malaise, weight loss and fever. An infected cat can spread the disease to other cats before it develops clinical symptoms itself. A blood test can detect whether a cat has been infected. Although a vaccine is available, the most reliable way to prevent feline leukemia is to prevent the cat from coming into contact with cats infected with the virus.

Feline infectious peritonitis (FIP) is an inflammation of the peritoneum (lining of the abdomen). Although FIP is contagious, some cats seem to develop a natural immunity. An infected cat may have no symptoms. Once the cat develops symptoms, the disease is invariably fatal. There is no reliable blood test for FIP but a vaccine is available today.

4.3 Vaccines

Cats can be successfully vaccinated against many serious diseases. Kittens should be vaccinated against rhinotracheitis, feline plague, feline enteritis and, optionally, feline pneumonia. Most veterinarians recommend a series of two to three vaccinations every three weeks, starting at three weeks of age. At twelve weeks the cat can be vaccinated against rabies (in countries where it is endemic), feline leukemia and infectious feline peritonitis. Vaccinations should be repeated annually to maintain immunity.

5 EXHIBITION AND VALUATION OF THE CAT

Many cat owners, even crossbreed cat owners, enjoy exhibiting their cats at shows or exhibitions.

Judges must be trained and accredited. Purebred cats are judged on health, temperament and because they meet the various characteristics of their model. Outcross cats are judged on health, temperament and general appearance. All cats must be docile and may be disqualified for biting or injuring the judge in any way.

5.1 Cat associations

A cat association is an organization where cats and kittens are registered, shows are organized and judges are selected. In most countries there are several cat associations. Cat clubs, breeders and owners who show their animals choose which association they want to join and which breed models they want to follow.

5.2 Cat shows

A growing number of local, regional and national cat shows take place throughout the year, with hundreds of cats competing for trophies. Owners show their cats for fun and to earn a reputation among breeders and other exhibitors. Shows rarely have cash prizes and entry fees and travel expenses can be high.

Although the exact rules and procedures vary from association to association, the forms are similar. There are usually four categories of competition: pedigreed kittens, pedigreed adults, pedigreed cats that have undergone surgery (neutered or without uterus or ovaries) and pets (crossbred cats or kittens).

A single cat show can have between 8 and 20 different judges and usually each cat is judged by all the judges. At many shows each judge has his or her own *ring, an* area consisting of 10 to 15 cages and a judging table. The cats wait in cages in another area of the show, called the show area. The owners, upon hearing their names, bring the cats into the *ring and* place them in the judging cages. The judge takes each cat out of its cage, places it on the table and carefully examines it to make sure it is healthy and conforms to its breed pattern. After judging all the cats of a specific breed or category, the judge awards preliminary prizes, for example for the best color or best cat of each breed. After judging them all, he gives the big trophies to the ten best in each category. Each judge works independently and their opinions may differ greatly.

6 POPULAR HISTORY OF CATS

Cats figure in the history of many nations, are the subject of superstitions and legends and a favorite motif of artists and writers.

6.1 History and legend

Cats were the object of worship in Egypt because of their ability to reduce the population of mice in the grain fields of the Nile, which was of paramount economic importance. The Egyptian goddess Bastet, represented with the body of a woman and the head of a cat, was the goddess of love and fertility. Cats were also a sport for the Egyptians; tied to leashes they hunted birds for the family table: the master would throw a boomerang that knocked down the birds so that the cat could pick them up and deliver them to the master. Because of their economic usefulness, and because they were believed to bestow many children, cats were so revered that they were sometimes mummified for burial with their masters or in tombs designed for that purpose.

Although Egyptian law forbade the removal of sacred cats from the country, Phoenician sailors smuggled them out. Cats were sold like other treasures from the East and, in ancient times, were found all along the Mediterranean coast. The Romans were apparently the first to introduce them to Europe.

The value of cats as predators was recognized in the middle of the 14th century, when a plague caused by rats, known as the Black Death, attacked the European population. Nevertheless, during the Middle Ages cats were hated and feared. Because of their nocturnal habits, it was believed that they had dealings with the devil. This association of the cat with witchcraft has been to blame for many acts of cruelty towards cats over the centuries. The Renaissance, however, was a golden age for cats. Almost everyone had one, from members of royal households and their servants to the peasantry.

In India, cats usually played an important role in religious and occult ceremonies. In South America, the Incas worshipped sacred cats, which are depicted in pre-Columbian works of art in Peru. Cats continue to be worshipped in countries such as Thailand and China.

6.2 Cats in art and literature

Egyptian tomb paintings and sculptures are the earliest depictions of the domestic cat. Images of cats appear on Greek coins from the 5th century BC; cats were later depicted in Roman mosaics and paintings, as well as on pottery, coins and shells. In the 8th century Irish manuscript *Book of Kells*, cats and kittens are depicted in one of its illustrations. Later artists, such as Leonardo da Vinci and Dürer, are among the many who included cats in their works.

Although the Old Testament makes no mention of cats, the Babylonian Talmud speaks of their admirable qualities and encourages the breeding of cats 'to help keep houses clean'. Memorable cats in literature include Rudyard Kipling's 'The Cat Who Walked Alone', the Cheshire Cat, the joint creation of English writer Lewis Carroll and illustrator Sir John Tenniel in the children's classic *Alice's Adventures in Wonderland* (1865) and Edgar Allan Poe's *The Black Cat*. Many contemporary comic strips and cartoons also feature feline characters that delight cat lovers of all ages.

RAISING A KITTEN: USEFUL TIPS AND TRICKS!

The time has finally come: your kitten can come home. And now you have an important task; time to start raising your kitten. It's all about enjoying the new member of the family, and it can be exciting. Use these parenting tips to give your kitten a healthy foundation for a safe and happy life and, of course, a good relationship with you.

Do you have everything you need to raise your kitten?

When raising a kitten, it is important to have a number of things at home. Think, for example, of baskets, toys, a brush, nail clippers and a scratching post. But that's not all, of course, take a look at our kitten needs checklist to get a complete list and determine what you need - good preparation is half the work!

Tips to kitten-proof your home

Just as with a baby or puppy, when raising a kitten it is important to prepare your home. Your kitten knows almost nothing about the world around it and has to learn step by step what is and is not allowed and you will want to avoid dangerous situations. When raising a kitten, it is advisable to kitten-proof your home. We are happy to help you with ideas on how to do this.

- **Kittens like to explore, so make sure they can't reach poisonous things.**
  Find out what things are poisonous to cats: plants, medicines, pesticides, cleaning products and foods (e.g. chocolate).
- **Make sure there are no things in the house that your kitten can easily swallow.**
  Think, for example, of coins, rubber bands, needles or toys.
- **Think ropes and loose wires.**
  Store them or secure them with suitable material. Make the house kitten-proof.
- **Store fragile objects in a safe place.**
  Do you have a nice statue on the windowsill? Chances are the kitty will accidentally drop it on the floor. If you can place

anything breakable in a place where your little adventurer can't reach it.

- **Make sure your kitty can't get caught in anything.** Think washing machine, dryer, curtains, tilt-and-turn windows, blind strips, etc.
- **Make sure the kitten cannot drink from the toilet or fall into it.**

Yes, even that is something to think about when raising a kitten....

- **Do not leave windows or doors open.**
  Consider also the balcony door and do not place the windows in a tilted position (there are special frames for tilted windows).
- **Do you have open stairs at home?**
  Don't let your kitty discover them until she's a little older.
- **Trash: Get a lockable trash can that your kitty can't climb or fall into.**
  Don't leave loose garbage bags containing waste. This smells very interesting, and your new friend may scatter garbage around the house or eat things that are not appropriate.
- **Is your kitten old enough to explore the garden?**
  Check for plants that are poisonous to cats.

Did you know: From the age of 5 weeks, a kitten's vision is fully developed. They can now see much better in the dark than humans, but they see fewer colors than we do.

Make sure your kitty does not play with loose strings or wires.

Help your kitty feel at home fast

When raising a kitten, it is very important that it feels comfortable. Your kitten will gain many new impressions during the transition. Saying goodbye to his littermates and mother is also a big change. Give your kitten a warm welcome and prepare your home with CatComfort. These products contain a copy of your own body's pheromones. The mother cat releases pheromones while nursing to create a sense of security and protection. With CatComfort you can mimic this feeling and can, for example, treat the litter box or travel basket with the Spray. Or place a Vaporizer_in the house to immediately indicate that this space is pleasant and safe. To get to know other cats it is best to use the Collar or Spray to encourage the relationship between cats. These tools are useful to make your kitty feel at home.

Before you start raising your kitten, it is a good idea to give it time to get used to its new home.

Kittens need a lot of sleep. It is important that you give it a quiet place to do so. For example, choose a room where no other family members (children and other pets) enter and let him get used to the new environment. Here you can put a bed, a litter box and a food and water dish. If you have children, let them know not to disturb while your kitten sleeps. Make sure there are enough (safe) hiding places in the rest of the house. Also, it's a good idea to put an extra litter box somewhere else (for example, on another floor).

By nature, cats like to climb. Make sure your kitty has opportunities to climb around the house. For example, with some boards or a large scratching post. This way your kitty will feel at ease anywhere in the house.

Did you know...? While nursing, the mother cat releases the mother pheromone. This pheromone strengthens the relationship between mother and kitten and provides a sense of security and protection.

Make sure your kitty has plenty of opportunities to climb, scratch and hide.

Tips for potty training your kitten

In general, kittens are fairly easily housebroken. This is because cats are very clean animals. In addition, the mother cat has usually already taken care of this part of kitten rearing. If your kitten still needs help, here are some tips:

- **Provide sufficient litter boxes.**
  This means at least one box per cat + one additional box.
- **Use the cat litter your kitten is used to from its time with the breeder.**
  This will familiarize him with it and make it easier for your kitten to use the litter box.

- **Place the litter box in the right place.**
  Make sure the box is in a quiet place that the kitten can easily reach. So not directly in a busy hallway, but also not somewhere far away in a corner on the third floor. Also, don't place the box too close to the food or water bowl.
- **Remove the lid of the litter box.**
  This will make your kitty feel more at ease.
- **Put your kitten in your litter box when you notice he needs to use it.**
- **Clean the litter box regularly.**
  For this purpose, preferably scoop out the dirt twice a day and change the cat litter every 1 to 2 weeks. In addition, clean the litter box regularly with warm water and a suitable cleaning agent.

Tip: Want to avoid unpleasant odors in the litter box? Beaphar odor neutralizer <u>absorbs</u> urine immediately, encapsulates the odor and breaks down the urine with natural bacteria.

Kittens can be tamed quite easily. Preferably use a litter box without a lid.

What food do I feed my kitten?

While raising your kitten, a good diet is also important. When the kitten has just moved into your home, it is best to feed it the food it is used to. If you want to change foods later, gradually mix the new food with the old one.

Choose high quality foods made specifically for kittens and ask your pet store which food is best for your kitten.

Raising a kitten: socialization is very important

When raising your kitten, good socialization is important so that it can grow into a confident and well-balanced cat. The first socialization period (age 3 to 8 weeks) takes place with the breeder. The second socialization period (from 9 to 14-16 weeks) takes place partially with the owner. During this period, the kittens are still very open to new experiences. After that, they are naturally more suspicious of new things.

It is very important during this period that the kitten comes into frequent contact with people. Pet your kitten regularly, give it treats in your lap, pick it up, cuddle it, play with it, etc. Let other family members do the same and regularly introduce your young cat to friends and family. An additional advantage of this is that, in addition to getting used to people, he will also build a good relationship with you.

Play and cuddle with your kitten regularly so that he learns to trust people and builds a good relationship with him.

Raising your kitten also includes introducing them to other pets. Do this gradually and give them time to get used to each other.

You may need to take him to the vet (e.g., for vaccinations). Make it a fun experience by giving him treats. Let your kitty get used to the travel basket first. Open it up in the room and put a nice pillow and something tasty in it. Make it fun and practice with the door closed once your kitty gets used to it. Gradually increase the duration. CatComfort spray can help make the travel basket a more familiar place. After your first visit to the vet, you can continue to practice getting her used to the travel basket to make the next time even more fun.

Finally, it is also a good idea to get your young cat used to sounds, such as fireworks, thunder or the vacuum cleaner. Play these sounds softly, increasing the volume little by little. Make it a good experience by playing with him or giving him some treats. You can get him used to the vacuum cleaner by turning it on briefly and rewarding your kitten. If you include this in your kitten's education, you'll be surprised how quickly he feels safe in his environment!

Gently introduce your kitten to other pets.

Training your kitten

Everyone knows that you can housebreak a puppy, but this is also possible with kittens! This is where the real raising of your kitten begins: it must learn what is and is not allowed. An important rule in kitten rearing is: don't punish, reward! When your kitten does something you don't want, it is important to keep in mind what you do want.

For example, does he climb on the countertop? Think of an alternative that fits you both. Often, the kitty is curious. Create a place where she can sit and still see the counter. This way, his curiosity is calmed and he still feels part of the group. When you have found that ideal spot, put a treat there every time you have to do something on the counter and reward him if he watches quietly from that spot. Also use the same principle when teaching your kitten to scratch on the scratching post, rather than on the furniture.

Reward your kitten for good behavior, so that it learns what is allowed and what is not.

You can also use a clicker as an aid. Want to know more about using a clicker? Contact a cat behaviorist.

Tip: Never teach your kitten to play (and bite) with your hands and feet. Divert attention from your hand to a toy that is appropriate.

If you follow these tips when raising your kitten, your sweet kitten will grow up to be a happy, well-adjusted and well-mannered member of the family.

TIPS FOR EDUCATING YOUR CAT AT HOME

An **educated cat** is one that lives in harmony with the rest of the individuals in a house. To achieve this purpose, it is necessary to understand the nature of the cat as a species and adapt its needs and ours. Cats are characterized for having an independent character and a marked territorial tendency; and it is something that as owners it is convenient to understand and respect to avoid problems with your new **pet**.

Free-roaming cats spend most of their time resting, grooming, hunting and leaving marks on their territory. For a cat to feel good when living at home, it needs to express these natural tendencies. It is not enough just to have its basic needs covered, such as food and water, it is also a requirement for the cat to be able to explore the environment, mark **its territory** or play at chasing objects simulating the act of hunting.

Our goal as owners is to find a balance for the development of the **cat's innate behaviors** within the home, achieving a relaxed and adapted individual, without being uncomfortable in coexistence.

How do I relate to my cat?

Socially the cat is defined as **independent**. The feral cat lives alone and does not create groups. It seeks the company of the opposite sex only for reproduction. At home, we should encourage its independence by letting the cat decide the degree of social contact it wants to maintain with each member of the family it lives with. Therefore, we should not force him to be with us or pet him longer than he considers acceptable. Besides, he likes to enjoy moments of privacy away from other cats or people.

Cats have a very accentuated **play behavior**, especially in their first months, and their favorite distraction is to chase moving objects as if they were hunting them. Therefore, if we play with them using our hands or feet we will get them to bite us. The ideal game is to use a paper ball or toys specially designed for them and play one or two sessions a day. Playing daily favors interaction with them and minimizes stress and, in addition, exercise prevents the appearance of certain diseases related to sedentary lifestyles and obesity.

Punishment should not be used because it is not an acceptable method of **education for the cat**, but a source of stress, fear and even aggression.

Keys for the cat to adapt and feel comfortable at home

Cats are very attached to the place where they live due to their territoriality, that is why it is necessary to take into account some issues so that they adapt to the home and be as comfortable and calm as possible. Therefore, one of the keys is to **avoid moving the cat.** It is very stressful for them to change places and that is why we should only move them when it is strictly necessary. It is better to leave the cat at home when we are absent for a few days and that someone comes daily to feed him and clean his tray.

In the same way, when **changes** are made **in the house** we modify the characteristics of the territory. This happens when we make renovations or buy new furniture. Obviously, sometimes these changes are necessary and to avoid the stress of our feline companion we can rub with a towel, previously used by the cat, on the new surface; for example: a carpet.

When we bring friends or relatives home, warn the people who are temporarily going to **live with the cat** that they should respect their space and not force the interaction, leaving it up to the cat to decide the time and duration of contact.

An important idea is that the cat must always have a place of escape or escape where he feels safe and from where he can contemplate his territory and everything that happens in it. Do not limit spaces with closed doors or force interaction with people or other pets. The approach to these should be gradual and with the option for the cat to escape from the situation to a safe place if the contact is not proving pleasant for him.

Preparing your cat for the arrival home of a baby or other pet

Another **stressful situation for the cat** is the arrival at home, its territory, of a new individual, either feline or human. The ideal with another cat is to make a gradual approach in which initially there is only olfactory contact and, little by little, as they tolerate each other, increase the contact until they can coexist in the same space.

The **cat's response to the arrival of a new baby** often causes concern for owners. The most important thing in this case is to encourage adaptation during pregnancy, try not to alter their physical territory or change their daily routine and let them smell the new clothes and investigate the new spaces created for the baby.

In the same way, the arrival of a new cat at home should be controlled and progressive, imitating natural behavior. In free-living cats, colonies of 20 or more individuals are formed and females seem to be more tolerant than males. When a new individual wants to integrate in that colony, it prowls the limits of the territory so that the cats of the established colony get used to its presence and especially to its scent. If all goes well, in a few days it will become part of the group.

The board

For all the situations described above, **synthetic feline pheromones** can be used, which are plugged in like electric air fresheners and emit cat pheromones, creating in the cat a state of well-being and placidity that minimizes stress.

How to stop cats from scratching all the furniture

Many owners, one of the things that bothers them the most, or directly puts them off having a cat as a pet, is the fact that they end up with all the furniture in the house scratched and the curtains and carpets torn to shreds? Well, we must understand that **scratching** is a natural behavior of the cat that has several functions, such as marking the territory, allowing the grooming of the claws and stretching the muscles. Therefore, inhibiting this action is a mistake that creates anxiety in the cat and can translate into behavioral problems later on.

To avoid that the behavior can be directed to inappropriate places such as doors, sofas or carpets, the ideal is to have a scratching post with characteristics and location that guarantee its use. This scratching post should have a stable base and a vertical scratching surface. It should **be placed in the cat's resting places** and, preferably, where it has scratched before.

A cat with sharp nails and a carpet is always a synonym of danger...

One of the most common mistakes is to place the scratching post in a hidden area, where aesthetically it does not bother us. This way, the cat will never use it and will continue to groom his nails on our wonderful leather sofa. These scratchers can be incorporated into interactive platforms designed for cats that provide a three-dimensional space, since cats enjoy jumping and climbing on high places.

A homemade alternative to commercial platforms is the placement of shelves where they can climb. Our cats enjoy high places because it allows them to have a three-dimensional view of their territory, as well as a place to escape or run away since many of them live with children or dogs that cannot reach these places.

How to teach a cat to urinate inside the litter tray

Cats are very clean animals, they spend part of the day washing and grooming their fur. From the time they are very young, they learn, in a practically innate way, to defecate in the litter tray. To **promote the use of the litter tray**, it is necessary to take into account its physical characteristics, the number of trays, the type of litter and its location.

At least two litter trays should be provided for each cat living in the house. The tray should have a wide, low rim and be large enough for the kitty to turn around. It should be placed in a quiet area, free of passage, accessible and away from electrical appliances. It is necessary to clean of excrement daily and change the litter weekly.

There are different **types of litter** on the market and each cat has its own preferences:

. Clumping litter is one of the most popular cat litters. This litter is characterized by forming a ball when it comes in contact with urine that is easily collected. This feature makes it easier to keep the litter box clean.
- Classic sepiolite litters create too much dust and are not recommended for cats with respiratory problems.
- Glass litter camouflages the odor very well but many cats do not like the feel of it.

Whole, unneutered males may **urinate outside the litter box** to mark territory. This urine is a marking urine and the cat will deposit it all over the house except in the litter box. This can be a problem for coexistence because this urine has a strong and unpleasant odor. In this sense, it is recommended to sterilize males that are not going to be used as breeders.

The cat should eat only its own food, avoid spoiling it with homemade treats.

Guidelines to educate and stimulate your cat with a healthy diet

If you have a new kitten at home, remember that it is important to give it a **diet appropriate** to its age, sex and health situation. In addition, cats drink very little water, due to their desert origin, so it is interesting to promote their consumption with drinking fountains, let them drink from the tap... It is better to put the drinking and feeding bowl in separate containers.

An interesting practice is to **hide food around the house** or use food-dispensing toys. The purpose of both is to encourage the search and hunting instinct. The cat is kept entertained in the countless hours we sometimes spend away from home due to the pace of life we lead and, in addition, we stimulate exercise in him.

To prevent the cat from stealing food from the garbage or the kitchen counter, we must educate it from the time it is a kitten to **eat only its own food**, avoiding homemade food and treats. Once the cat becomes an adult, it fixes its food predilections and it is difficult for it to feel interest in trying new foods.

The cat's nutritional needs will depend on its age: puppy, adult or senior and its health status. Many feline diseases, such as diabetes, obesity or the presence of crystals in the urine, can be prevented or combated with specially formulated feeds.

BIBLIOGRAPHY .

Alvarado, R. 1970. *Los Felinos: gatos, leones, tigres, leopardos...*Editorial Noger, S.A. Barcelona-Madrid, 80 pp.

Bephar. 2024. *Raising a kitten, Useful tips and tricks. Pets.*

Animal education. 2015. *Tips for educating your cat at home.*

Gibbon, D. and J. Burton. 1976. *Man's faithful friends: dogs, cats and horses.* Creaciones especializadas de Artes Gráficas, S. A. Barcelona, 125 pp.

http//www, wikipedia. com

Microsoft Encarta. 2009. *Cat.* 1993-2008 Microsoft Corporation. *http//www. Encarta.com*

More sources
    Web Links

    Pets
    Website dedicated to pets. Dogs, cats, fish... Offers related links and information on the care and prevention of pet diseases. In Spanish.
    http://www.mascotas.com/
    Friendly pets
    Website with detailed information for the good care of pets.
    http://www.mascotamigas.com/
    My animals
    Extensive information on all types of pets.
    http://www.misanimales.com/

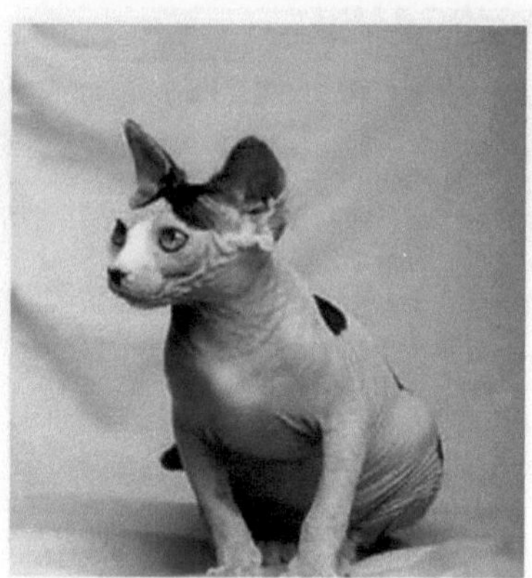

Printed by Books on Demand GmbH, Norderstedt / Germany